Unexpected item
in the bagging area

Unexpected item in the bagging area

Chris Martin

Michael O'Mara Books Limited

First published in Great Britain in 2012 by
Michael O'Mara Books Limited
9 Lion Yard
Tremadoc Road
London SW4 7NQ

A CIP catalogue record for this book is available from the British Library.

Papers used by Michael O'Mara Books Limited are natural, recyclable
products made from wood grown in sustainable forests.
The manufacturing processes conform to the environmental regulations
of the country of origin.

ISBN: 978-1-84317-944-3 in hardback print format
ISBN: 978-1-84317-983-2 in EPub format
ISBN: 978-1-84317-984-9 in Mobipocket format

1 2 3 4 5 6 7 8 9 10

Cartoons by Ian Baker
Designed and typeset by Design 23

Printed and bound in Great Britain by Clays Ltd, St Ives plc

www.mombooks.com

Contents

SHOP 'TIL YOU DROP

Cheap plastic toys

In their ongoing attempt to destabilize the West, our Chinese overlords have unleashed a torrent of plastic toys and dolls. These smiling, wall-eyed monstrosities crowd the shelves of our toy shops, supermarket checkouts and museum gift shops like a malevolent invasion force from some distant psychedelic planet.

When you were growing up you had virtually no toys. A lone action man, a beloved Barbie doll, maybe an Etch A Sketch or a space hopper, all made in Hong Kong and all cherished because you knew that breaking them meant no toys until your next birthday. These longtime companions have been replaced in our children's affections with a rotating gallery of faceless, bleeping bedfellows.

Your kids can barely find their pyjamas for the number of cheap plastic toys clogging up their

bedrooms. You give your kids toys for being good and to stop them being bad. They get toys in their cereal packets, with their Happy Meals, as mementos of family trips out and to while away rainy afternoons. On birthdays and Christmas your friends join in to create a tottering mound of cheap gifts that take your children most of the day to open.

While these toys play an eminently disposable role in your kids' lives, they are absolutely indestructible in the real world. In 1992, an armada of twenty-nine thousand plastic yellow ducks, blue turtles and green frogs made in China for the US firm The First Years Inc, were washed overboard in the eastern Pacific. Since then they have travelled seventeen thousand miles, some landing in Hawaii and others spending several years frozen in the Arctic ice. They have floated over the site where the Titanic sank and even outlasted the ship which was originally carrying them.

With this in mind, every parent can look forward to finding disembodied Lego men and random Moshi Monsters in suit pockets and handbags for many years to come.

Coffee shops

Coffee shops are everywhere. You can't turn a corner without seeing a Starbucks, a Coffee Republic, a Costa or a Caffè Nero, often right next to one another and all staffed by cheery students with baseball caps, nose rings and burns all over their forearms.

Such is the ubiquity of coffee shops that they've been forced to seek new real estate, erecting them in hospitals, petrol stations, even a job centre. In fact it's only a matter of time before they start building them on the only free ground remaining – inside other coffee shops.

Big, huge or vast

You can remember a time when coffee came in one size and was either black or white. Now you have to decide if you fancy an espresso, a mocha or latte and whether it's tall (big), grandi (huge) or venti (vast). You will also need to decide whether you want a heart-starting extra shot or heart-stopping whipped cream. Also on offer

are marshmallows and a dash from a selection of tempting bottles stacked behind the counter that look like booze but taste like cough syrup.

Cardboard bucket

When you finally hand over the money, the trouble really starts. The resulting beverage is so large it needs to be served either in a cardboard bucket or a china basin which takes two hands to lift. You can only wonder what Samuel Pepys would have said.

> **'Win or lose, we go shopping after the election.'**
> Imelda Marcos, former first lady of the Philippines now more famous for her collection of 2,700 pairs of shoes.

Organic food

Anyone who has had any exposure to modern children knows they have gone badly wrong. All you can remember when you were a child is bland food, uninspiring toys, early bedtimes and an all-pervasive fear of adults. Your kids treat the world with the unchecked self-regard and wilful abuse of their fellow man more commonly found in the palaces of Middle Eastern dictators. Desperate for an explanation, you convince yourself that the borderline psychotic behaviour manifested by your offspring must be caused by something in their turkey twizzlers. You decide to go organic.

As nature intended

For food to be certified organic, it must be produced without the use of synthetic pesticides, fertilizers, antibiotics, or food additives – and as you'd rather these chemicals were used to clean your kitchen rather than the inside of your stomach this sounds good. The organic movement further claims food grown as nature intended has benefits, both for the environment and personal health and even that it has higher levels of minerals than its non-organic counterparts. Again – it sounds great. Finally, organic food is not irradiated. In practical terms this means fresh food will probably go off before you get a chance to cook it. In short, you'll be spending a lot of time in your local organic superstore.

Mom and Pop Organic

The organic superstore looks pretty much like a normal supermarket if you didn't bother to paint the fittings and tripled the prices. But what really grates is that every single product on the shelves has a story printed on the back. Usually this is how old-timers Mom & Pop Organic decide to knit a range of toilet paper on their small holding or, more accurately, how Old Jed Wholeearth's pasta sauce tasted so good that he demolished his bankrupt farm and built a factory on the site to produce it in industrial quantities using tomatoes imported from China.

Despite the fact that you hold homicidal feelings toward every organic producer you read about, you persevere for the good of the family. But as you wander the shelves browsing the brown rice and tofu, you notice something odd. Your fellow shoppers are made up of smug middle class couples, stick thin health freaks that look close to death and ageing banker's wives wearing jewellery worth more than your car. There's not a child in sight. This is because organic food, far from pacifying your kids, sends them into an uncontrollable rage the moment they realize that all the chocolate in the place has been replaced with carob and the closest they'll be getting to a packet of crisps is a dry oat cake. The subsequent in-store meltdown necessitates an immediate and permanent retreat to Tesco.

Decline in standards

To add insult to injury, an independent study published in 2009 by the Food Standards Agency assessed evidence from the past fifty years of nutrient levels in crops and livestock and found that eating organic food offered no health benefits at all.

Hollister

Every week your kids add another retail destination to the family shopping trip. You've come to accept that the likes of Abercrombie & Fitch, Billabong, Urban Outfitters, and The Gap exist solely to drain the current account of parents across the land, and this would be all right, if your teenagers' clothes weren't twice as expensive as yours.

Endless summer

The popularity of these shops is beyond you. They all seem to sell the same thing: sportswear designed for an endless Californian summer, when your kids live in a rain-drenched suburb and the nearest body of water is so polluted that they'd grow an extra toe if they stepped in it.

But the real genius of these brands is not the clothes but the shops in which they are sold, and most monstrous of all is the Hollister shop. Situated in the middle of a busy shopping mall that you didn't want to go to in the first place, and looking from the outside like an alpine ski lodge, there is no more unfriendly environment on earth to the middle-aged parent than the inside of a Hollister shop.

The dark room

It is pitch black and throbs with such unspeakably loud music your fillings rattle. As you stumble across the threshold you are confronted not by clothes – because you can't see any – but by an indifferent but wildly attractive Lolita in a bra top, or an enthusiastic but shirtless would-be surfer with abs like a xylophone. In a split second you realize that you will pay anything just to get out of there and into the relative calm of the electrical retailer. Your kids know this and so do the marketing department at Hollister.

> **'We used to build civilizations.
> Now we build shopping malls.'**
> Bill Bryson, writer and Anglophile

Supermarkets

When you were growing up there was only one supermarket – now called a superstore. It was in the centre of town and everybody went to it for their weekly shop. Then the big grocery chains, spurred on by hundreds of thousands of pounds of market research and a desire for unchecked expansion, started messing with the size of their stores leaving you feeling more like Alice in Wonderland than a hungry punter. There's probably method in this madness but what do you know? You're just a customer who's trying to buy a loaf of bread.

Use your loaf

First you try the hypermarket, essentially the old superstore, but massively increased in size and cut adrift in a new location far outside town and marooned in a sea of parking. It sells things that you never previously associated with the supermarket like TVs, kitchen utensils and sports equipment. Arriving at the hypermarket you become lost in the aisles as you search for your seeded batch. Overwhelmed by the sheer volume of stuff on sale, you panic buy enough canned goods to keep you going for a year.

Daily express

Driving back from the hypermarket with the boot of your car clinking like an off-licence in an earthquake, you notice a new mini 'Express' supermarket has opened up on the high street. You wonder why you bothered to drive all the way out to the hypermarket in the first place.

This new store sells most of things you need for the week, only at higher prices than either the hypermarket or the superstore but – as they don't sell everything you need – you still have to go back to the hypermarket anyway. And you still can't buy a loaf of bread as the shelves which would have sold them have been replaced with an in-store microbakery specializing in glazed rolls and heavily-thumbed croissants.

Charity muggers

Every time you open your email you are confronted by an announcement that yet another of your paunchy, middle-aged friends is attempting a breathtaking feat of stupidity using the notional excuse of raising money for charity. While it might be their choice to freeze to death on the side of Kilimanjaro or suffer a heart attack on the nineteenth mile of the Copenhagen Marathon, the bottom line is that you are going to have to fund their midlife crisis. While you can just about bring yourself to pledge money to enable this madness, what really gets your blood up is that your money will be going to yet another charity which you have never heard of and have no interest in.

Questions, questions

Whenever the government issues some new stealth tax, we are outraged. Questions are raised at the highest levels and government ministers are duly grilled on TV. But the charities regularly fleece us at every turn and no one says a word.

Virtually every time you buy a drink or a snack, a proportion of the sale goes to build sports fields or put computers in schools; if the charities don't get to you on the shelves, there are buckets for your change by the tills. You are asked to buy plastic flowers, badges and key-rings to pin to your lapel to show support for dead soldiers and live orphans. And where the utility companies have had the good grace to stop cramming your letterbox with paper bills, the charities have taken up the slack with an unending torrent of unsolicited mail shots featuring sour-faced, one-eyed children and grinning cancer survivors all demanding their slice of your hard-earned cash. All this comes on top of the direct debit that you foolishly signed up to as a student to help raise a child in Tibet, which is still being claimed, despite the child now having children, if not grandchildren, of their own.

Stand and deliver

With the charities putting so much time and effort into finding new and innovative ways to siphon off your money, it would come as no surprise if they decided that it would be quicker and easier

to stop you in the street and demand that you turn out the contents of your pockets … which of course they already do. You can barely walk across a railway platform, through a subway or across a shopping mall without being leapt upon by a pensioner in a sash shaking a plastic goblet or a cheery student with a clipboard and sheaf of direct debit forms.

Really it's all a matter of balance. And the balance is so much in favour of the charities that you have come to the conclusion that it might be easier to sign over your entire pay check to them in the first place, leaving you free to conduct a series of sponsored moonwalks and skydives to raise the necessary funds to buy food and groceries. It would certainly be more fun.

> 'It is only when they go wrong that machines remind you how powerful they are.'
> Clive James, writer and raconteur

Unexpected item in the bagging area

A machine should be more efficient at serving us in a shop than a human being, especially when that human being is a heavily-pierced teenager in an ill-fitting nylon uniform with a cider hangover. In practice, this is not the case because the person operating that machine is you.

The weakest link
The fact that you are the weak link in this chain is the last thing on your mind as you jump the long supermarket queue and confidently bowl up to the automated check-out. All you will be doing is taking neatly-packaged items from your

basket of groceries, passing them in front of a laser scanner then placing them in a plastic bag. It's something you've watched the human cashiers do a million times with the kind of 'please kill me' look in their eyes that can only mean the task is far from challenging.

The Da Vinci bar code

However, when you get going, it's a different story. None of your chosen items have the bar code in the same place. You spend five minutes turning each item over to find where to scan. When you finally locate the bar code, some will scan easily with a satisfying beep, while others require careful smoothing of the wrapper or to be held at the correct angle. Then you pick a bunch of bananas. These have no bar code at all and need to be weighed. You scroll through an endless series of menus more reminiscent of finding a document on the office intranet than a visit to the shops.

You give up and look plaintively around the shop for help. The heavily pierced teen – who has now served the four people who were originally in front of you in the queue – trudges around the counter and assists you by nonchalantly using one hand to key a secret code at incredible speed.

Things look like they are on the up until a foolish attempt to get another bag from the stand results in the computer announcing there is an unexpected item in the bagging area. To the designers of these machines this phrase may mean something but to you – who doesn't know what the item is or where officially constitutes the bagging area – it is one of the most oblique string of words in the language.

Waiting for the man

What is clear is that you can't scan any more items until it's sorted which means waiting for the return of the teen who has managed to serve another three customers in the interim. Before you can splutter a series of excuses, the teen has swiftly and efficiently reset the machine and – without making eye contact – run through the rest of your groceries at the same time.

Red-faced and flustered, you hustle out of the shop and resolve to order your groceries online in future.

I'M ON MY MOBILE

The ringtone draw

There are three kinds of mobile phone owners: people like you who get a new phone and leave it exactly as it is, people who get a new phone and fiddle with the settings to change the ringtone, and then there are the downloaders of ringtones. This last group also create their own wallpapers, customize its vibrations and put the thing in a garish novelty case with the result that their mobile looks more like a child's craft project than a communication device.

Most parents use pictures of their children as a screen saver on their phone but when the kids turn thirteen this has to stop: a forty-five-year-old man with a photograph of a smiling fifteen-year-old in a bathing costume on his phone will find himself in a police cell long before he can explain that it's an innocent picture of his daughter on a family holiday.

While you have nothing against those who wish to personalize their phones, you accepted a long time ago that you are too lazy and too inept to change the ringtone on your phone. You got your first mobile back when they were relatively rare and primarily owned by drug dealers, rent boys and the middle management with abusive bosses who were prepared to pay outrageous call charges to harangue their staff in the middle of the night. In those glorious days it was novelty ringtones that made the veins on your neck bulge but as mobiles have become ubiquitous it is the out-of-the-box ringtone that has become the scourge of your day.

Is it for me?

Despite Nokia, HTC, Apple etc. offering us more ways to personalize our phones most of us have no inclination to do so. As a result, everyone has the same ringtone. So when a phone rings on any train or in a meeting, the chances are it could be an incoming call for any one of four people sitting within a five-yard radius of one another. This is fantastic when the call is for you but unbelievably humiliating when you're the one who eagerly grabs their phone to discover it's not.

Despite the fact that everyone perks up like a gang of meerkats when a phone starts ringing, no one will risk reaching for their phone for fear of embarrassment. You find yourself in a Mexican stand off, unsure whether you should be the first to go for your phone and risk looking like a fool. It goes without saying that by the time someone finally cracks the caller has rung off.

'Technological progress has merely provided us with more efficient means for going backwards.'
Aldous Huxley, author of *Brave New World*

Music on mobiles

It is the duty of the young to annoy the old. While the dress sense of today's teens is annoying – trousers worn beneath the buttocks (boys) or horrible velvet tracksuits (girls) – it pales into insignificance against their love of listening to music on their mobile phones.

Terrible noise

If you happen to travel on a bus or train during daylight hours, the chances are there will be at least one group of under-eighteens listening to dubstep through the woefully inadequate single speaker of a mobile phone. The music itself is terrible and seems to consist of an incessant hissing more reminiscent of a malfunctioning kettle, interspersed with repetitive bleeps and a digitally-enhanced singer. The lyrics, such as they are, extoll the virtues of driving a Mercedes for an audience whose only source of income is pocket money.

The horribly tinny sound is not a concern to

young people. Thanks to a condition known as presbycusis the ability to hear high frequencies deteriorates in people over twenty-five. This condition was used to devastating effect in 2008 with the launch of the Mosquito. This was an electronic device that could be used to deter loitering young people by emitting a sound of approximately 17.4 kHz – a frequency that is ear-piercing to them but undetectable to their elders and betters.

Know your rights

There was an outcry about its use from the muesli-eating Left, with the pressure group Liberty going so far as to suggest that the device may violate the Human Rights Act of 1998. Seemingly no one aged over thirty-five agreed because when Baroness Sarah Ludford MEP suggested a motion to ban the use of the Mosquito, it failed to get sufficient signatures from the adults in the European Parliament to proceed to a full debate. Indeed, you quite fancy the idea of installing the device to deter juvenile loitering in your own home, possibly around the fridge, telephone and television.

Unfortunately, even the application of

revolutionary technology did not stop young people from being annoying. They actively embraced the Mosquito and turned it into a mobile phone ringtone known as 'Teen Buzz'. Because it couldn't be heard by teachers, 'Teen Buzz' was the ideal ringtone to use during classes.

'Using Twitter for literate communication is about as likely as firing up a CB radio and hearing some guy recite *The Iliad*.'
Bruce Sterling, science fiction writer and journalist

New smartphone

The phone companies issued a newer, shinier and sexier handset a couple of weeks after you got your last one so yours ended up looking like an obsolete brick that even Thomas Edison would have been ashamed to be seen with. It's been a long, painful, contractually designated twenty-four-month wait until you could upgrade to something state of the art.

Beam me up

Despite the fact that your old phone works perfectly well, you're pretty confident that investing in a wafer-thin, glass-and-metal slab which looks like a prop from *Star Trek* will impress work colleagues and the opposite sex alike with just how digitally savvy you are.

However as soon as you pop the cellophane on its gleaming white box and your pornographically exciting new gizmo drops effortlessly in your palm, it dawns on you that everything else you own is designed to destroy it. House keys will score great canyons into its screen, half-eaten boiled sweets will clog up its nano grills, pocket fluff will bespoil its sheen. While it won't fit into the pocket of your jeans, it will fit neatly into the mouth of a child or small dog. Even your sweaty fingers leave unsightly greasy marks on the graphite.

Protect and survive

So it's back to the mobile phone shop you go to spend another fifty pounds on cases, covers and screen guards which make your lovely new phone look like something you picked up cheap at a Tupperware party.

'A squirrel dying in front of your house may be more relevant to your interests right now than people dying in Africa.'
Mark Zuckerberg, founder of Facebook

TECHNO WARS

Drone attacks

Of all the recent applications of technology, surely the unmanned aerial vehicle (UAV) or drone is the most extraordinary. There is something wonderfully reassuring about silent automatons dishing out discreet aerial death to faceless enemies thousands of miles away. Sort of like someone taking the rubbish out for you as you sleep at night. Just as the internet has kept us out of the supermarket and satnav tells us where we are, the drone takes the hassle out of the nasty business of duffing up foreigners.

Remote controllers

All of which would be great if the drones were not being controlled by a group of recently enlisted school leavers in a giant call centre in Nevada. It will only be a matter of time before some hungover and heartbroken twenty-year-old employs this same technology to provide him with twenty-four-hour surveillance of his ex-girlfriend's house or to incinerate the car

of his former headmaster. Only then will we realize that basing our military strategy around a group of pimply youths who have spent the last five years in their bedroom playing *World of Warcraft* may not have been that sensible.

> **'I haven't reported my missing credit card to the police because whoever stole it is spending less than my wife.'**
> Ilie Nastase, tennis ace and three time divorcee

Syncing gadgets

Ever since the invention of the stacked hi-fi system in the 1970s, geeks have been working to splice every piece of technology together.

In the heady days of the hi-fi stack, this dismal practice was kept in check by practical limitations – notably that the cassette deck had to be no more than a cable's length away from the tuner which made syncing an exercise in piling things on top of one another. However the invention of Wi-Fi has meant that the gloves are off.

What Wi-Fi?

Wi-Fi has allowed every gadget in your home to talk to every other gadget. The result is a technological landscape in which you wouldn't dream of buying a toaster unless you could also order books from Amazon on it.

Chief among the offenders is your phone. There was a time when phones were safely attached to the wall by a cable or sat on a special table along with the phone book. Now they

follow you everywhere. Your phone syncs to the computer, it docks nicely in the top of your CD player, it tells half a dozen programs where you are and you can use it to operate your TV and access your email. If you have an iPhone, you can even use it to discover that you've missed your wife's birthday.

Fit for purpose

You begin to wonder where you fit in. You're not synced to anything and are starting to feel pretty redundant. Maybe you should skive off to the cinema and let your phone go into the office alone. It could probably do a better job.

iPad

In the final quarter of 2011 Apple sold 15.4 million iPads, doubling the company's profits from the same period in the previous year to a cool £8.54 billion. While these booming sales figures are impressive, to you they only confirm one thing: you simply must buy an iPad.

Go west

Like everyone else in Western society, you are a tech-addicted consumer who believes that buying anything square and shiny with an ominous screen will vastly improve your standing among your friends, make you more efficient at work and generally give a sense of meaning and purpose to the hollow farce that is your life. It might even contribute to educating your kids because God knows you have neither the time nor the inclination to do so.

But one question lingers at the back of your mind as you stride into the Apple store with your credit card held aloft: what does an iPad do?

Is the iPad really a big iPhone? Or is it just a computer without a keyboard? Is it for watching films or reading books? It's clearly too big to replace your mobile yet, it's much too small to replace your TV. You can't really type on it, nor can you make phone calls, so it's no good for business. It doesn't have a joystick or a keypad so it's really no good for games.

Cool stuff

While it would clearly be great for chopping vegetables on or as a drinks tray, they never seem to be used for that in the adverts.

Of course, these doubts don't stop you handing over the cash and scuttling home like a squirrel that's burgled a nut factory. When you get home it soon becomes clear that all the iPad is really good for is looking at cool stuff – by which you mean looking at your own house from above on Google Maps. Is that really worth four hundred quid? Of course it is.

Kindle

You love to read so it comes as something of a surprise that you don't do it anymore. The only time you buy books in a bookshop is at Christmas and even then it is usually a last-minute purchase for a difficult-to-buy-for nephew.

Determined to get back on the literary ladder, you pay a visit to your local bookshop. It has closed down, so you get back in the car and finally find a bookshop buried in a nearby mall. Stepping across the threshold you are quite surprised to find that half the shop floor is taken up with gifts, novelty bookmarks, CDs, children's toys and a coffee shop. There are a couple of books at the back so you accept this diversification in retail bookselling as another sign of the times.

As you browse this meagre offering you realize that, despite seeing yourself as a keen reader, most of your discussions about books at dinner parties are based entirely on reviews

you read in the Sunday supplement. If you are honest, ninety per cent of your actual reading takes the form of celebrity gossip in free newspapers on the way home from work, recipes and idly flipping through twenty-year-old paperbacks while sitting on the loo. You resolve to rectify this situation immediately by buying a new Kindle.

Read all about it

Superficially, the Kindle is pretty exciting. It comes in a nice box and looks a bit like an iPad. The manufacturer claims it is really easy to search for and download books and you duly install twenty free classic titles from *Jane Eyre* to *Dracula* that you've always wanted to read but never did. But you soon discover that the Kindle successfully recreates what you didn't like about books in the first instance: they take ages to read. In a time when you can play basic games on the colour display on your fridge, this situation is too boring to be tolerated. Admittedly, you can do some clever things like add bookmarks and make a note on the text but you never did that with real books so what's the point? To add insult to injury, when you download new digital titles they're barely any cheaper than the heavily discounted hard copies in the shop.

In the end, rather than saving you money, the Kindle turns out to be a more expensive way to read because, two weeks later, you leave it on the train.

> **'If you change lines, the one you just left will start to move faster than the one you are now in.'**
> O'Brian's Law

Apps

You have a lot of apps. Apps for recipes, apps for your bank, apps to play games, apps to read books and newspapers.

There are now more than half a million apps available via the iTunes store and by the look of your monthly bill from Apple, you have downloaded most of them.

The excitement builds

Periodically you scroll through pages and pages of these things on your iPhone and start to wonder what they all do. You can remember being tremendously excited as you downloaded the app on the recommendation of a friend, whose friend was equally enthusiastic when they demonstrated how invaluable it was. It is true that you can pop open an app to see when the next bus is coming, or to navigate a 3D image of The Louvre or to find out what's going on at NASA but – if you're honest – the need has never arisen.

It was when you downloaded a word processing app that you had the sneaking suspicion that apps only exist because Apple can't be bothered to install any half-decent software on their products. Whether it was the iPhone, iPad or MacBook, you were so pant-wettingly excited to unwrap your new toy that you never stopped to notice that it came with barely enough pre-installed software to change the clock settings.

Sales figures

Apple is perfectly happy to leave you to download and pay for the software needed to run your tech separately; not least as they've earned an additional $4.9 billion in cumulative sales since launching the App Store in 2008.

What is going to happen if this app culture spreads even further? Perhaps we'll get used to cars where the drivers have to kneel in front of the steering wheel because they are too skint to purchase the windscreen, doors and seats.

'In department stores, so much kitchen equipment is bought indiscriminately by people who just come in for men's underwear.'
Julia Child, America's original television chef

DOWN
TIME

Razor wars

People have been shaving for a long time and they've always hated it. Although seventy-five per cent of men and eleven per cent of women do it every day, it ranks up there with a visit to the dentist as something most people would rather avoid.

Archaeologists have discovered that early man originally used sharpened bones and sea shells to smooth their faces until somewhere around 3000 BC when Egyptian priests created a razor specifically for the purpose. These early razors were fashioned from bronze so we can be fairly confident they were an improvement on hacking at your face with an oyster. This fact was not lost upon the subsequent generations of men and women and the razor industry is now worth billions.

Despite the invention of the razor, shaving remains time-consuming, bloody and painful so why do the razor companies make so much money? The answer is that the marketing men have realized

that the application of state-of-the-art technology to what is effectively a very small knife, combined with the promise of pain-free depilation means big bucks.

When you buy a new razor you must navigate a minefield of confusing additional features, from lubrastrips to vibrating handles, all scientifically proven to deliver a closer, smoother shave. But that's nothing compared to the unstoppable arms race of adding extra blades.

For nearly five millennia, razors had a single blade that everyone shaved with and no one complained. Then some bright spark at Gillette came up with the idea of a dual blade razor because – logically – two blades must be better than one. Competing companies soon began to add additional blades to their razors and now you find yourself contemplating whether you should go online to specifically purchase the Pace Shave, a razor that boasts a lethal looking trimming blade on the back and an utterly pointless array of six – yes, six – blades on the front. You're lucky they didn't throw a couple more on the handle for luck.

Faced with a razor that resembles a solar panel on a stick, that single blade razor starts to look pretty good.

Cosmetic consumerism

You like to think that you're pretty content with the way you look, even though you've always held a grudging respect for those who are prepared to go under the knife to achieve a straight nose, a smooth jaw or whopping pair of double Ds. You like to think that you are not so vain that you would feel the need to have surgery and ignore the fact that you lack the necessary courage to take the plunge anyway. Instead you choose to plough a considerable amount of your time and hefty proportion of your disposable income into any product that promises radical improvements to your looks without recourse to a stay in a private clinic.

This goes a long way to explain why you can't shut your bathroom cupboard for bottles of fake tan, facial rehydration lotion and skin firming

gel. You know these potions are no better for you than rubbing lard on your skin on a hot day. You're a sucker for anything that smells of aloe vera and is advertised by either a rugged man rubbing his unshaven chin or a youthful-looking, recently showered woman, particularly if there is a pseudo-scientific animation to go with it.

But just how far can you go to look good without being vain? Is having your teeth whitened or electrolysis on your hairy bits as bad as Botox? And does having laser eye surgery or unflattering braces on your teeth count as plastic surgery or a commitment to long-term health?

Before you lose too much sleep over this, it might be worth considering that it doesn't really matter if you have smooth tanned skin and a dazzling smile, when your paunch is hanging over the top of your designer jeans like your bottom half has melted.

New telly

It's time to get a new TV. The old TV is not broken; on the contrary, it works fine. But a visit to a friend's house where there are flat screens in the kitchen, the bedrooms and something in front of the sofa that looks more like an IMAX cinema tips the balance in favour of an upgrade.

For a long time, the only thing a new telly had to offer was a minor increase in size. Every three years, you effectively bought the same product but with a screen that was an inch bigger on either side. This natural expansion was kept in check by the fact that every inch the TV manufacturers added at the sides translated into four inches at the back and about a stone in weight. Unless you lived in a long thin room with a reinforced floor eventually you'd be in trouble. Then they went flat screen. Without a vast hot box behind them TVs could become massive and still leave you with plenty of leg room.

Then the egg heads looked for new ways to keep us glued to the box. So it is that you find

yourself in the electrical shop being shown a 3D HD LED sixty-five inch TV with full internet access by an enthusiastic middle-aged man with bad breath and a clip-on tie. When combined with four sets of powered 3D glasses, eight surround speakers and a wall mounting unit, this Smart TV is more expensive than a Smart Car. You put it on the credit card and arrange for an industrial courier to ship the thing to your home.

Cut to three weeks later. One wall of your living room is taken up by the daunting TV, but you're up in your office watching *The Apprentice* on your iPad. It turns out that the new TV has such disturbing pinpoint clarity that you are frightened to turn it on for fear of having to endure every pore on the lead actor's face. Explosions in surround sound set off the car alarm in the street outside and it uses so much electricity that when you change channels, the streetlights dim.

Sizing of clothes

You don't want harp on about size zero models or the fashion pages being dominated by morose, anorexic teens but – to be blunt – if you are killing time with a fashion magazine, you would rather that the people in it were young and thin.

While you might expect unchecked body dysmorphia among the people who create, style and exhibit high fashion on the catwalks of London, Paris and New York, you despair when it creeps into the manufacture of the workaday, humdrum rags you buy from suburban chain outlets to clothe yourself. You are not looking for a signature piece from Galliano's Spring collection, just a top and some jeans that will be comfortable and smart enough to wear to work on dress-down Friday.

The first hurdle you face is that in no two shops is the sizing of trousers the same. A thirty-four waist in one shop will be so tight you feel like you are being cut in half and so loose in another it looks like you're wearing clown pants.

And when you do find a pair of trousers that fits your waist the legs will be so long they should come with a complimentary pair of stilts. The situation is as bad with shirts, although if you have long baboon arms and a neck like a lolly stick you should be all right. As a quick look around the shop reveals dozens of podgy middle-aged people with stumpy limbs and generous rumps struggling to recapture their youth and not a single fourteen-foot tall, stick-thin giant with arms that drag on the ground, you have to wonder who the hell they are making these clothes for.

The ultimate blame lies with us. Our refusal to pay more for our clothes relies on thousands of malnourished teenagers working long hours in some sweatshop in Asia. It seems safe to assume that these third world workers have little insight into our world of potbellies, ready-meal inflated thighs and poor posture honed by hours in front of the goggle box. Either that or where they come from the idea of a shirt worn so tightly that your man boobs protrude from your front like a pair of googly eyes is considered attractive.

Newspapers

For the past decade, we have listened to doom and gloom stories about how mainstream media is failing and that soon newspapers will be a thing of the past.

As the media is in such financial peril, it is ironic that you can barely move for newspapers. There is a daily version found in newsagents and smaller-sized versions of the same thing for airports. There's an online version, a mobile version for your phone and a tablet version for the iPad. And that's just the broadsheets. Try travelling on public transport on any day of the week and you'll find virtually every seat buried under so many discarded copies of *free newspapers* that you are forced to sit in a newsprint nest.

Story time

You don't really read the newspaper any more. They are full of stories about children being eaten by Rottweilers, celebrity marriages and

endless stories about the decline in newspaper ad revenue. They are simply too depressing to bother with.

Arguably, the real reason why newspapers are in trouble is that some bright spark thought it would be a good idea to give them away. Whilst you don't claim to be the world's greatest economist, even you can see that employing people to write a newspaper, printing the copies and delivering them all over the country, must incur considerable costs, so that handing them out like curry house flyers is daft.

> **'It has become appallingly obvious that our technology has exceeded our humanity.'**
> Albert Einstein, Noble Prize winning physicist

Café in the park

When you were a child, the café in the park was up there with hospital bins and open sewage overflows as one of the premier urban biohazards. A failing greasy spoon that was part urinal and part unsanctioned needle exchange,even parents as disinterested and incompetent as yours knew not to take their kids there.

But the café in the park has undergone something of a renaissance. Nowadays it's a deli with grassed play area, nice wooden tables and a queue of sleep-deprived yummy mummies in outsized designer shades snaking out the door. It's all because some bright spark realized that the trick to drawing in the middle-class punters was not cheap and cheerful fare with free corkage on super strength lagers, but launching a wilfully expensive eatery that would make the prices in a Michelin-starred restaurant seem reasonable.

With such a captive audience, it is surprising no one worked this out sooner. If you're a parent with young kids you will spend up to fifty per cent of your life in the park. The prospect of spending four hours pushing swings and feigning interest in rabid-looking squirrels fills you with such horror that the ten minutes reprieve offered by spending fifteen pounds on a coffee and two pastries for the kids seems like a bargain.

This being the case, why do the shark-eyed restaurateurs even bother with nice food? Come rain or shine, you will be in the park, traipsing around after your pretentiously named offspring who behave like a bunch of mental patients after a smash-and-grab raid on a scooter shop.

You could argue that middle-class parents demand a higher quality of al fresco fare but the real reason for the growth of this culinary highway robbery is much simpler – the promise of health. If you want to keep your spouse off your back, you can't take the kids to McDonald's. Besides, what's the point of the café owners installing an expensive deep fat fryer when they can charge you five pounds for an organic cheese sandwich with no butter?

Cold hard cash

For the most part, you would be happy to see the back of cash. Carrying enough cash to get through the day is a hassle. A trip to the funfair with the kids – while never an exercise in prudent financial planning – now requires a roll of notes that could choke a horse and you still have to drop by the ATM on the way home if you want to pick up a pizza.

Cash is becoming obsolete

But the debit card system presents problems of it own. It lacks the geezer glamour of a fat wad of notes or the jet-set feel of a platinum credit card. Your debit card has been sat on, left in the sun, and had tea spilt on it; as a result, it looks more like a tramp's passport than your key to financial independence.

But debit cards can be useful when shopping. Putting aside the nerve-racking few minutes you have to endure as you keep entering your date of birth rather than your pin number, and the

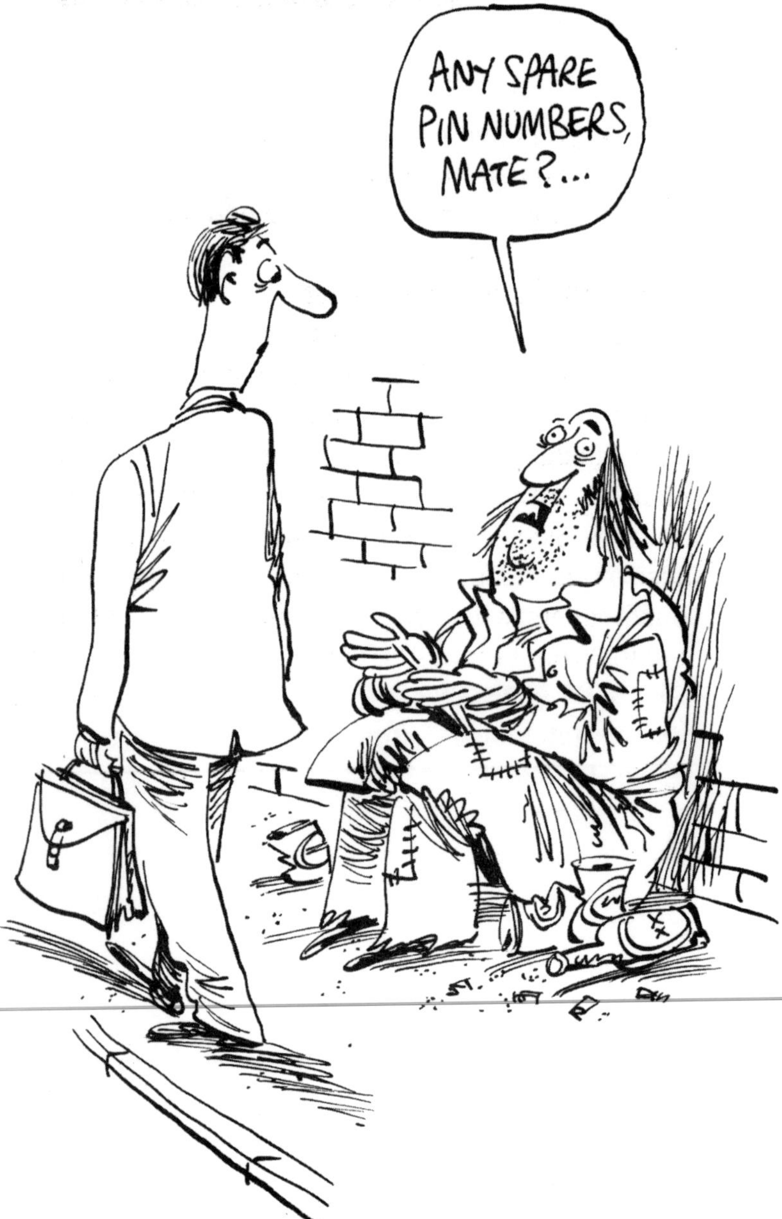
ANY SPARE PIN NUMBERS, MATE?...

fact that you have no idea how much the goods you purchased actually cost you, paying by debit card is a lot easier than counting coppers out.

Pennies from heaven

As long as there are children, there will always be cash. While you relish having a chance to finally get rid of the fourteen tonnes of loose change that is clogging up a vase in your hall, your kids will not thank you for it. They keep themselves in bus fares, sweets and iTunes vouchers by riffling these coin caches (as well as going through your pockets when you're drunk and scooping pennies from behind the sofa cushions).

Nike+

For years you never took any exercise and you never put on any weight. Instead you lived on overindulgent sandwiches and rivers of booze. But now it's time to face the fact that you're not getting any younger.

Every pizza and takeaway curry adds another couple of centimetres to your spare tyre. A night on the sauce leaves you so puffy-eyed and hungover you look like you're in anaphylactic shock. So you make the decision to get fit.

Less is more

A bit of research reveals that if you want to lose weight all you have to do is eat less and do some exercise. This seems simplistic, so it's off to the sports shop, where you are delighted to discover that a whole world of high tech science has been applied to the tricky business of creating your very own training and personal development programme. In short, you can spend the afternoon shopping rather than going for a run.

After an hour of baffling pseudo-scientific nonsense rattled off by a young man in a tracksuit with a neck like a fat thumb, you emerge from the shop five hundred pounds lighter, the proud owner of running shoes that look like props from a sci-fi movie and a humiliatingly tight Lycra outfit.

Sofa surfing

In a rare moment of self-awareness you realize that you're unlikely to get off the sofa without prompting. You can't afford a personal trainer so you invest in a new iPod and a Nike+ system. The Nike+ system consists of a piezoelectric sensor that you put in a special hole in the sole of your trainers and a receiver that tracks and analyses your journey to fitness. Thus equipped, you can enjoy runs in which the music on your iPod is constantly interrupted by a disturbingly upbeat American voice counting down the kilometres, (you can't work out how to switch it to miles), until you hit your goal – usually the newsagents at the end of the street. While this disembodied encouragement is undoubtedly motivating as you jog confidently out of the house, it starts to grate when you are bent double, pouring sweat and fighting for breath half a mile later.

Even worse, you have foolishly chosen to connect your Nike+ system to both the Nike website and your Facebook page in the belief that this will somehow keep you on track. This means that the distance of your runs and the time it has taken you are now posted online; so your friends can follow your enthusiastic first attempts then watch you get slower and slower over shorter distances, before finally stopping all together.

> **'Some things man was never meant to know. For everything else, there's Google.'**
> Author unknown

TiVo

If there's one thing you don't need in your life, it is more television. So it comes as something of a surprise that some well-placed advertising persuades you to buy a TiVo box.

Despite the hype, TiVo turns out to be just another DVR box that lets you record TV on a digital hard drive. In the US TiVo is as synonymous with DVR television as Hoover is with vacuuming. There it makes a lot of sense to be able to pick and choose what you watch when your cable service offers you ten thousand channels, half of which are in a different language.

Shine on

Nonetheless, a new shiny object is a new shiny object and you are hopping from foot to foot when the engineer arrives with your new box. You're delighted to see that the TiVo box comes in menacing black, has some exciting lights and an impressively complex remote control.

But after spending a taxing evening working out how to get CNN on it, you make a terrible discovery. The opportunity to record whatever you want, whenever you want, has highlighted just how little you actually want to record anything. You like some of the soaps but do you really want to commit to recording every episode including the Sunday omnibus, for the rest of time? You probably should record the latest high-definition scientific opus, but as you've fallen asleep halfway through every one to date is it really worth it?

Make no mistake, you do watch a lot of TV and you love it more than life itself, but when you have to justify choosing one show over another rather than just letting the soothing cathode rays wash over you in a brainwashing mist of contentment, you draw a complete blank. Besides, you can get the news on the internet and half of the shows on the cable channels are repeated with such regularity that there is no need to record them anyway.

'I don't shop because I need something, I just shop for shopping's sake.'
Cat Deeley, television presenter

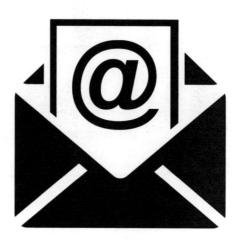

I'M ONLINE

Spam email

The unwanted mail that used to pour through your letterbox was simple to understand: gas bills, party invitations, and – if you bought something from a catalogue – yet another catalogue offering products of interest solely to retired people with the time to ponder buying a collapsible snow shovel for the car. All this junk meant that you had to shoulder barge the door open when you got home from holiday. Still, you got to use the recycling bags the council dropped off.

Declutter

Email, far from replacing all this clutter, has actually increased it. The gas company now feels obliged to email you a note that your bill is coming, a reminder that your payment is overdue and – somewhat optimistically – a customer satisfaction survey, along with a personal note from the managing director explaining their commitment to customer service (and higher

prices). In place of the party invitation, your friends expect you to commit to parties on Facebook or, even worse, to endure a fourteen-minute, poorly-animated, digital card on your birthday.

Tick the box

Some spam is bearable. We can live with the odd offer of a foolproof way to lengthen a manhood or a wonder drug promising unlimited sexual prowess. But forget to tick the box to unsubscribe from email updates when you buy something online and you'll be receiving emails packed with last minute offers from the florists for the rest of time.

And guess what? It's not easy to read things on the screen so you'll be printing them out as well.

> 'The greatest thing about Facebook, is that you can quote something and totally make up the source.'
> George Washington

Internet shopping

We all live busy lives. Between working longer hours, hitting the gym to justify the huge annual membership fee, dropping the kids off at school and finding time to plough through thousands of hours of *The Killing*, there's barely a moment to get to the shops.

Thank God for the internet

Not only is it great for essentials, it's also fabulous for buying stuff you never knew you wanted or needed. Whether you're sourcing original vinyl, baby booties from French artisans or a polystyrene coffin full of frozen steaks from an organic farm in Scotland, a world of unnecessary purchases is at your fingertips. In fact, you now spend so much of your evenings picking out reclaimed tiles on eBay that you barely have time to wonder when and where all this stuff is getting delivered.

It isn't. At least, not to your house. All you'll receive in return for your hours in the front of the screen is a series of illegibly signed cards pushed through your letterbox by a delivery man informing you that your hard-researched purchases are available for collection from an industrial estate five miles away.

> **'To err is human, but to really foul things up requires a computer.'**
> Author unknown

Google Maps with Street View

At the touch of button you can use Google Maps to see streets and houses all over the world, the north and south poles and even the surface of the moon.

For someone as poor at directions as you, the whole thing has been a boon. No longer are you late for meetings, as you wander aimlessly up and down some strange road looking for a client's ambiguously-named office block. You've also discovered that it is great for scoping out if the hotel you've booked online is actually wedged between a brothel and dogs' home before you get there. However, not everyone shares your enthusiasm.

Public inconvenience

A man in the village of Angers, France is suing Google for ten thousand Euros, arguing that the image the technology giant put online of him urinating in his back yard violated his right to privacy and ruined his reputation in the community. But taking a whiz in your back yard is tame by Street View standards. Google has also captured hundreds of bizarre images such as an apparently naked man emerging from the boot of a car, male and female couples snogging, crimes in progress and a host of men furtively entering 'adult entertainment' stores.

All a blur

While Google routinely blurs the faces of people captured by their Street View cameras, it can be argued that clothing, local knowledge and the fact that they are relieving themselves in the garden of a house they own makes them pretty recognisable.

Google Street View is a bit like the local newspaper. You secretly harbour a desire to appear in it and wave coyly as a Google car goes by.

You just wish they had given you adequate notice before they took a picture of your house. When you call up your beloved homestead on Street View, you are embarrassed by what you see. The bins are overflowing, the front garden is alive with weeds, the windows hang open like a teenager's mouth and the whole place needs a paint job. Thanks to pin-sharp digital photography the house you pay a fortune to the mortgage company for looks like it's one up from a squat. No wonder your friends never come to visit.

> **'I have never met anyone who wanted to save the world without my financial support.'**
> Robert Brault

Password protection

It's hard to know which faceless technocrat decided it was a good idea to password protect everything that you might reasonably want to access in the course of your day. But nowadays your laptop, mobile phone, office network, gas bill and your spouse are held behind military-grade, 256-bit encryption. While it is reassuring that hackers can't access your overdraft, it is absolutely infuriating that you can't either.

Random characters

Most of us struggle to recall our own postcode so it seems crazy that we're expected to deal with baffling combinations of letters, numbers and random characters issued by everyone from our car insurers to the supermarket.

As a result we've all stopped caring. According to Mark Burnett's book *Perfect Passwords:*

Selection, Protection, Authentication, the world's most used password is '123456'. Number two on the list is simply 'password'.

Remember you're a Womble

The solution is surely to give us just one password that works for everything. In fact your parents should select this for you at birth and insist it be tattooed – backwards so it can be read in a mirror - on your forehead. And if walking around with 'womblefiend123' on your face in Gothic script seems a high price to pay, remember that at least you'll be able to check your bank balance.

> 'Why does the internet always seems way more interesting when you have work to do?'
> Author unknown

Social media

In 2009, Nielsen reported that the internet had seen a seven hundred per cent increase in the use of social media. What was really surprising about the report was that this growth was fuelled not by tech-obsessed youngsters but by two groups that really should know better: thirty-five to forty-five-year olds and those over sixty.

The basic offering of social media is keeping in touch and sharing your experiences. This makes some sense when applied to those under thirty-five who are getting out to bars and clubs, dating new people and going on exciting backpacking holidays but how does it work for you? You spend the majority of your time at work and the rest slumped in front of the telly. You never go out and when you do it's only to attend a dinner party with people you've spent so much time with you could draw a map detailing every line on their ageing faces.

This lack of activity hasn't stopped you getting heavily involved in social media. You dutifully signed up for accounts with Twitter, Facebook, Foursquare and Instagram; installed the apps on your phone and spent a heady evening making connections with friends, friends of friends, family members and, in some cases, people's children and pets.

You are tremendously excited by the sense of purpose setting up these accounts has given to your humdrum existence, but you find that you hit a block when it comes to actually sharing information because a life in which the highlight of the week is ordering a Chinese takeaway from the Golden Panda on a Friday night is hardly worth reporting.

Strictly old school

Luckily, your new best friends seem happy to supply the solution to this problem. Despite the fact that they're always in when you call and are usually watching the same TV shows as you are, on social media they have managed to post the evidence of a glittering existence of weddings, gigs, old school reunions and laughter-filled family outings. It isn't long before you stop trying to post the embarrassing truth about your own life and devote yourself to living vicariously

through their online adventures.

On social media, everything looks more vital and exciting than it actually is. Even the events that you attended in person, when reflected through the prism of other people's opinion, are wildly improved. A night which you remember as being defined by someone spilling red wine down your front, awful food and your best friend confiding in you that they were thinking of getting divorced, looks like a kaleidoscopic joyfest on Facebook. Furious minute-by-minute reporting of key points at conferences by tweeting colleagues seem like a much better version of events than you sitting at the back of the conference hall wondering when the next coffee break is.

An informal survey posted by Forrester in 2012 revealed that fifty-six per cent of people felt that the time they spent on social media was wasted. You would beg to differ. Unless things start getting more interesting in your life soon, social media may be all that stands between you and suicide.

Call of Duty

Once a trip to your local record shop promised a relaxing afternoon nostalgically browsing through discounted CDs of classic albums and arty DVDs; now it offers nothing but a minor panic attack when you find yourself lost between row upon row of garishly packaged video games.

It's hard to think of anything more disruptive to one's sanity and sleeping patterns than video games. While playing them is bad, it pales into insignificance against the horror of living with someone who does.

Of these digitised tributes to mindlessly obsessive behaviour, none is more irritating to the sane person than *Call of Duty*. For nearly fifteen years, *Call of Duty* and its many unimaginatively titled sequels (*Call of Duty 2, Call of Duty 3* etc.) have gripped every male under the age of forty with the kind of fervour previously reserved for Premier League football and busty females.

Call of Duty is a first-person shooter, which

means that the player experiences the action via a set of disembodied hands holding a gun which pumps an endless stream of bullets into screaming Japanese soldiers, screaming Japanese zombies, screaming Nazis, screaming Nazi zombies or – latterly – mercifully mute terrorists. For anyone living with the player, this virtual blood bath will be experienced solely through a nerve-shattering cacophony of groans, explosions and muted gunfire coming from the next room.

In 2010 Activision, the makers of this tribute to testosterone-fuelled madness, announced that the franchise has already sold 55 million copies and made some $3 billion. Something must be done. Those 55 million shipped units mean that there are at least 55 million deeply unhappy spouses, parents and female siblings sitting on the sofa with their hands over their ears trying to get on the telly to watch *The X Factor*.

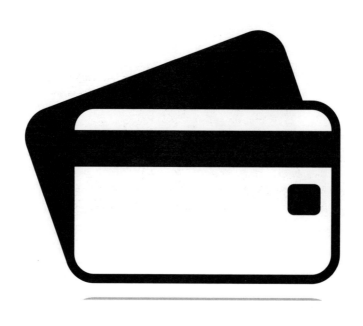

MONEY MATTERS

Loyalty card

It's official. You are addicted to loyalty cards but have no idea how many points you have accumulated or what you could do with them.

This fact hasn't stopped you filling your wallet with dozens of these things and dutifully producing them every time you buy groceries, clothes, coffee, or petrol. The brochure they came with seemed to indicate that using the cards was something to do with sun-drenched holidays or getting a new car but you can't be sure, as you didn't read the small print.

Points mean prizes

Every penny you spend with a loyalty card equates to a large and opaque number of points – 1754.56 for example – and there are different scales for cash purchases, debit card purchases, goods on sale and so on. So if you want to know when you qualify for the new car, you will need a decent grasp of theoretical mathematics to work it out.

Like any addiction, your unflinching loyalty to your loyalty cards will eventually drive you insane as you attempt to satisfy your need for points. It will start when you are consumed with rage when you discover your partner has done a weekly shop without using the loyalty card. Then you will find yourself regularly driving around for an extra forty minutes to find the nearest supermarket rather than risk missing out on vital pointage. Finally, you will hit rock bottom and find yourself 'tilling' – the despicable practice of picking up the discarded receipts of others to add the points to your own card.

The only answer is to admit that you are hopelessly enslaved to providing marketing data to the supermarket, cut up your cards and take it one day at a time.

Unexpected item in the bagging area

'I have come to the conclusion that
Google must be female, as she has
the answer to everything!'
Author unkown

Property porn

Tim Berners-Lee and the idealistic nerds at CERN who invented the World Wide Web envisaged some new global academic utopia. But as soon as their creation fell into the grubby paws of the general public, it became little more than a vessel for us to indulge our basest desires in the privacy of our own homes. At first these indulgences involved little more than ogling naked ladies (men) and bitching about how fat our old school friends look on Facebook (women), but now they have now turned to something far more sinister.

Gone fishing

Property porn is the debasing act of trawling the websites of local estate agents ogling the living rooms of houses which are just round the corner from yours (and slightly more expensive). Can there be anything more daunting to a man than coming home to find his partner and her friends, half-cut on mid-range white wine, gathered around the laptop and complaining loudly about the impractical floor plan of a house three doors away that has recently come on the market.

Like all addictions, your obsession with property porn is a guilty secret that starts small and soon escalates. It begins as you innocently browse a small flat as an investment property. But within months it has spiralled into avarice-fuelled, all-night sessions browsing vast suburban mansions the purchase of which could only be funded by a lottery win. Realistically you'd be better off sticking to hot-handed sessions of real porn. At least they have a logical end after which you can go to bed.

Extras in cars

The reality of your motoring life is three sweaty hours a day spent in a cramped tin box mired in rush hour traffic. You pass some of this time listening to drivetime radio and eating mints and the rest you spend drooling over the supercars trapped in the lane next to you.

This seething envy has been stoked by a hideous marriage of glossy television advertising and affordable part-exchange programmes. These have fed a fantasy of cruising along the open road behind the wheel of a sparkling new motor of your own.

Your current car is a disgrace. Several owners ago, it was desirable, even cool but now its once gleaming paintwork bears the scars of a thousand parking mishaps and the interior is the world's largest wastepaper basket.

When you get to the showroom, it becomes apparent that the car you saw in the advert, with its twitching LED displays, stainless steel gear knob and acres of heritage walnut panelling,

had been tricked out with 'extras' which push the price up to that of a starter home. Your bog-standard new car will have no sunroof, seats that look like they were recycled from a bus and an AM radio that can only pick up Hindi dance music stations. With 'extras', your new car could look and feel like something driven by James Bond; on your budget, it actually ends up looking like something driven by Noddy.

Forecourt is forewarned

To add insult to injury, any new car will lose between fifteen and twenty per cent of its value in the first year of ownership. From the moment your tyres leave the dealer's forecourt, you lose money as surely as if you'd incinerated a pile of tenners in his office to celebrate signing the papers.

Financial products

We're constantly told that the financial sector is vital. Similarly, we're also told that the whole thing is about to explode like a vigorously shaken can of coke that's been left in the freezer over night.

Like most people, you try to keep abreast of the issues of the day. As a result, you are vaguely and constantly worried about the financial crisis. Depending on who you talk to at various dinner-parties, the financial sector is worth between eight and fifteen per cent of our national GDP. Despite not knowing what this means, you are aware that it is quite a lot.

Too big to fail
We have been told by the government that the banks are too big to fail, but it has been made equally clear that they are quite comfortable

with the bank of 'you' failing on an epic scale that will leave you fighting foxes in street for discarded chicken bones to feed the family. In an effort to avoid this, you attempt to understand the many financial products you have been sold over the years.

You have insurance for your house, life, mortgage, car and for random acts of personal injury as well as a pension and a mortgage, yet you never got past the first page when it came to reading the small print. God knows what was in the equally obtuse contract you signed.

Out of focus

The issue is one of focus, and how you lost it in the deep boredom induced by the discussion of anything financial. Financial products are always advertised by beaming, healthy-looking families who seem to be able to parley a single trip to the bank into a deep sense of personal satisfaction and some kind of glamorous beach holiday. Yet every time you are sold a financial product it is by a shifty-looking twenty-year-old with greased-down hair and an ill-fitting grey suit who drones unconvincingly about repayment schedules. As you fill in the endless forms, you try to focus on the small print but you soon zone out and develop an incessant ringing in your ears which announces the early stages of a migraine. All things being equal, signing the papers and getting out of whatever grey faceless office you find yourself trapped in seems more of a priority than dealing with the sub-prime situation.

'Men are from Mars. Women are from Venus. Computers are from hell.'
Author unknown

Cold callers

Nowadays we have it pretty easy on the shopping front. You can simply log on to the internet and buy all the groceries, clothes, books and insurance you need at any time of the day or night. So are you really sitting at home waiting for a pushy teenager in a call centre in Pontypridd to ring you about great discounts on double glazing or whether you want to take part in a survey about your utility supplier? And even if you were, would you choose to have this person call during your favourite soap or halfway through dinner?

Fresh air

And the phone is only the half of it. Every time the doorbell rings after six o'clock it's a young offender with a tray full of overpriced dish cloths or a monosyllabic student trying to sign you up to a direct debit to bankroll eye operations in a developing nation. By the end of the evening it's a breath of fresh air to open the door to find

one of your neighbours tearfully confessing that they've run over your cat while parking their car.

But surely the true inconvenience of cold callers is that no one quite knows just how rude you can be to these people. In your dreams you slam down the phone or casually shoulder the door shut with a curt 'not now, you reptile' but such honesty is socially unacceptable. Instead, you are forced to waste valuable TV time concocting an unconvincing and convoluted excuse about running a bath or the children's bedtime, which inevitably results in you agreeing to speak to the cold caller at the same time the following day.

'Computers must be male. As soon as you commit to one you realize that if you had waited a little longer, you could have obtained a better model. In order to get their attention, you have to turn them on.'
Author unknown

ON YOUR TRAVELS

The quiet carriage

Travelling on the train is horrible. Assuming you can work out a ticketing system that is so complex it feels like the train company hired cryptographers to design it, you still have to face cancellations, delays and navigating a station forecourt that looks like a refugee camp.

When you finally get on the train, you find them so clogged with discarded fast-food wrappers and bottles that the carriages look more like the aftermath of a teenage sleepover than a modern transportation system. The seats are small and the less said about the toilets the better.

The most irritating thing about modern train travel is the constant and mindless use of mobile devices by the passengers. The most taxing hazard you used to face on the train was accidentally sitting down opposite a table full of squaddies making their way through a mountain of canned lager. Now, as soon as the train pulls out of the station, everyone takes out their mobile and rabbits on about being on the train and what time they're arriving, as if this information is of national importance.

But it doesn't end there. They kill time by obsessively checking their emails or their Twitter and Facebook accounts, before giving up and calling their office again to spin out an awkward and unnecessary conversation with a hapless co-worker. All this is accompanied by the usual nerve-rattling chorus of bleeps, boings and bings that you might expect when several dozen smartphones, dongle-enabled laptops, iPads and handheld games consoles are collected together in a confined space.

To seek some relief, you try to position yourself in the quiet carriage. Supposedly an oasis of peace, this is right at the far end of the train which means you have to sprint the length of the platform to make it. Far from being an oasis of peace, the quiet carriage is deafeningly loud. The autistic owners of mobile technology now find themselves at a loose end and talk loudly about how they really must make a call, as soon as they arrive, to their fellow travellers, the conductor and – if it comes to it – themselves.

Low-cost airlines

Like a rain-sodden barbeque or mosquito bite on your eyelid, a miserable trip on a low-cost airline has become a regular feature of your summer. The boom in low-cost air travel created a corresponding boom in the number of chavs, numpties, children and pensioners, clogging up our airports. Air travel has changed from a glamorous cosmopolitan escape to the transport equivalent of a long weekend at a holiday camp.

Bargain hunter

You love to go abroad but you also love a bargain. So as you book your summer holidays you put aside all the terrible experiences you have previously had with low-cost airlines, resolutely ignore your partner's pleadings and allow yourself to be seduced once again by the offer of flights for spare change.

The first thing you discover is that unless you are planning to travel in two years' time, the flight actually costs much more. And then you start to add the extras ... If you want to bring a bag (or your kids), book a seat, arrive at your destination earlier than midnight or pay with a credit card, you have to pay more. As boarding a low-cost airline is a high-speed scramble reminiscent of the Pamplona bull run you might like to pay for early boarding (effectively a head start) too.

Early riser

But it doesn't end there. As a matter of policy, all planes operated by low-cost airlines run late so you will spend at least an hour waiting in a hideously decorated Day-Glo holding area before boarding. As you had to get up at 5.30 am to check-in, you'll be pretty hungry by the time you get on the plane. Naturally there's no complimentary food on board so you can look forward to paying Michelin-starred prices for food which resembles something lurking around the back of the bins at KFC. The indifferent, demoralised staff deliver it to your seat then attempt to sell you a lottery ticket.

You will arrive at your economy villa hot, hungry, three hundred quid lighter and clutching a piece of paper which promises to return your bags from wherever in the world they have ultimately been delivered. At this point you promise yourself never to take a low-cost airline again … until next year that is.

> **'Airplane travel is nature's way of making you look like your passport photo.'**
> Al Gore, environmental activist

Automated check-in

When it comes to travelling the friendly skies, people are fast going out of fashion in the customer service department. It's possible to book and pay for a flight online without talking to anyone, but now you also have to print off a boarding pass, drive yourself to the airport and check yourself and your bags in using an automated terminal, before following further instructions on a screen to get you through the departures lounge and onto the plane.

Panic stations

On paper, this sounds like a dream but in fact it's a nightmare. For a start, the booking process is filled with jeopardy. You still struggle with the difference between 12 am and 12 pm, a fundamental confusion that has repeatedly turned disastrous when combined with the time difference at your destination. More anxiety ensues when you remember that you forgot to steal some printer paper from work so you will have to stop at a Kall Kwik to print out your boarding pass.

When you finally park at a short stay car park that costs more than your actual ticket and is situated three miles from the terminal, you are already cutting it fine. As a result you pass a tense fifteen minutes looking at your watch as you wait for the shuttle bus.

Still worrying about whether you're going to be landing at midnight the same day or lunchtime the following day, you arrive at the terminal to be greeted by a confused mob wandering aimlessly around the departures hall like something from a George A Romero movie. Some are staring blankly at the machines wondering whether they should be using their boarding pass or their passport to check-in, while others are stabbing wildly at the touch screen and arguing with their partner over which seat number is the bulkhead and whether they will have to pay more for it. A separate, more desperate group has detached itself and is trying to deposit their bags. This means alternating between the information desk and magazine store because they are the only places where they can see a human being in uniform.

You make it into the departure lounge and queue up for a coffee. It is then that you notice on the monitor that your flight has already been called, even though it is not leaving imminently. The reason for this is made clear by a small sign next to the monitor warning you that your designated gate is a twenty-five minute walk away. Unless you run, spilling boiling hot coffee down your front as you go, the plane will be gone by the time you arrive.

It's pretty clear that all this hassle could be averted if there was a nice girl in a beret and blazer to tell you where to go, but airlines have removed them all in favour of a sequence of processes, automated notifications and oblique signs worthy of *The Da Vinci Code*. Ironically the one place they have kept a uniformed person – the cockpit – is the one place they don't need one. Modern airliners have had the ability to fly on autopilot from one airport to another without the need for human intervention since the 1970s. Shame they can't say the same about the passengers.

'What does it mean to pre-board? Do you get on before you get on?'
George Carlin, comedian

Parking meters

Cars are great for getting from place to place but they are useless if you can't stop at your destination and get out. Unfortunately, armed with the triple excuse of health, safety and increased traffic flow, your council is hellbent on making this a reality. If you are forced to pull over because your back seat is on fire, you will be spotted on camera and fined remotely.

And if you are considering pulling into a parking bay, think again. As part of a plan to make all convenient services less convenient in the name of convenience, parking bays became an active fine hazard as soon as their meters stopped taking real money.

Coin drop

That's right; you now need a mobile phone, a debit card and an online account with a sinister automated payment bot to park your car. So rather than dropping a few coins in the meter and popping into the shops, you discover that you have to spend ten minutes walking up and down the road trying to find the bay reference number, then another ten minutes trying to remember your car's registration number, before working out that it is written on the number plate.

All this nets the council a tidy sum. But if you were planning on giving up, parking the car in your back garden and taking the bus, you can think again. They don't take cash anymore either.

'A real patriot is the fellow who gets a parking ticket and rejoices that the system is working.'
Bill Vaughan

The weather

In February 2012, the BBC weather map confidently reported that the temperature just outside Bangor in North Wales would drop to a frosty -99. Despite the fact that a temperature this low would have seen birds falling out of the sky and blood freezing in the veins of even the hardiest Welshman, no one batted an eyelid.

Take a rain check

This is odd because like most people you now check the weather online every few hours; even more so if you are thinking of stepping outside or, God forbid, actually travelling somewhere. In fact, you now check the weather as often as you check the time. This would be okay if it wasn't accompanied by general weather related anxiety: Are you wearing the right shoes in case it rains? Is this coat too heavy for the temperature this afternoon? Should you wait until a week next Tuesday when there is only a 19% chance of precipitation before you leave the house?

And the weather forecast is now everywhere. It is on your phone when you switch it on, it is your internet home page, it is displayed at train stations, on billboards – wherever you look there is a temperature gauge and half a sun. The only way to escape the weather is just to stay inside.

STUCK
AT
WORK

Email at work

Email used to be great. It promised a rosy future for your admin, the end of the tottering piles of paper which turned your desk into a walled A4 fortress and the beginning of a paper-free digital utopia in which you would finally catch up with your correspondence. But your colleagues had other ideas.

Freed from the need to print and deliver documents, or even lift up the phone to talk to you, your fellow workers have gone into email overdrive. Rather than wander the four feet across to your desk, they send two or three emails as a precursor to a meeting request (which you naturally ignore), followed by a half a dozen follow up comments CC'ed to everyone from the MD to the cleaner – most of whom feel obliged to reply.

Masterpieces

Don't expect these emails to be literary masterworks either. Sent on a whim, over-

familiar and insultingly casual (Hi there!), they are abbreviated to the point of incoherence and represent little more than the last gasp of their failing synapses.

A survey published by The Radicati Group in 2010 indicated that the average business user receives seventy-seven emails a day and sends thirty-nine. It did not indicate how many of these were to announce the arrival of the sandwich man in reception.

Adjustable orthopaedic chair

The time we spend at work is frighteningly similar to the time we spend at home. Both involve sitting on our widening bottoms staring at a screen, shuffling through an endless list of vaguely irritating emails.

To relieve the stress of your humdrum existence, and the agonising pain in your lower back caused by fourteen hours a day of immobility and biscuits, you will probably find yourself investing in an orthopaedic chair.

The orthopaedic chair looks much like any other chair but it has three key differences:

- It costs five times more than a normal chair.

- It has a bizarre arrangement of knobs,

levers, buttons and stalks all of which seem to do the same thing or nothing at all.

• It is designed to look impressive in a showroom but not to be sat on.

Getting on and off your new orthopaedic chair without sustaining an injury is a concern. A brief jiggling of the knobs makes it clear that – height-wise – you're looking at either squatting on the ground with your knees round your ears or finding a stepladder to get onto it. The angled seat and towering arm rests mean that when you finally get into the chair, you will be pitched forward with your arms stretched limply before you, giving you the appearance of a pantomime ghoul haunting your own PC.

On the plus side, your back feels great. Not least because the chair is so uncomfortable you now spend half of your day wandering round the office looking for a sofa.

The office kitchen

When you are trying to tot up a spreadsheet or plough through the morning's email, there a few things more off-putting than the stink wafting out of the office kitchen. On investigation this terrible smell turns out not to be a problem with the sewerage system, but one of your colleagues heating up a homemade concoction in the microwave.

It seems extraordinary when every available square metre of shop frontage around your office is filled with coffee shops, bagel bars and sandwich drop in centres, that you should need a kitchen at work. Yet a growing group of Tupperware gourmands insist upon it.

You only use the kitchen once a fortnight for reluctantly making your fellow team members a cup of instant coffee; and only then because the online rota shamed you into action. Your fellow office drones seem to view their activity in the communal kitchen as an unofficial audition for *Masterchef*.

Street market

Every morning you see people emerging with individual cafetieres of Fairtrade Colombian coffee and bowls of organic oatmeal. But it is around lunchtime that it really gets busy as these amateur chefs take an hour out of their allegedly busy days to slice vegetables, make curry, toss salad and boil noodles, converting what was meant to be a functional area for providing complementary tea and coffee to visiting suppliers into a scene more reminiscent of a Mumbai street market.

But what really grinds your gears is that although your colleagues have spent a great deal of time and effort, in and out of the office, preparing this impromptu trip through the world's cuisine, you can be sure they won't be cleaning up after themselves.

By five-thirty, the office is deserted and you venture into the kitchen to get a glass of water. The place looks like it has been vandalised. Crusty plates and dirty cups overflow from the sink, while every spare surface is covered with cooling pools of soup and half-read copies of the newspaper. The fridges fart open to reveal rows of carefully labelled plastic boxes, odd-shaped tinfoil parcels and half-drunk bottles of soya milk, while the floor of the microwave is coated with a gelatinous puddle of an undefined substance close to achieving sentience. Indeed it is so filthy that you wonder how much Thai green curry actually made it out of the microwave and into your colleague's greedy faces.

There is of course no way to beat them but – as you don't fancy getting food poisoning – you won't be joining them. You open the window next to your desk and pray the cleaners come early.

'Stress is when you wake up
screaming and you realize you
haven't fallen asleep yet.'
Author unkown

LISTED

Top 10 infuriating helplines

When calling a financial services helpline, the average person listens to nineteen minutes of recorded messages before getting through to a human being. As no company would dream of launching a new service today without an associated helpline, we can all look forward to spending a lot more time hanging on the telephone.

Tax office

Interminably engaged, followed by forty-five minutes of music before being cut off.

•

Doctor

A forty-five minute wait before being told to call an ambulance.

Gas company

Usually quick to answer but then you're told that your details are on the 'other system' which is – of course – not working.

•

Bank

Incorrect selection of options results in a twenty-five minute wait before being transferred to another department for another twenty minute wait.

•

Your office

A ten-minute wait before being transferred to your own voicemail.

•

The cinema ticket booking line

A five-minute call spun out to take forty-five minutes by menus and options delivered in a slow drawl.

Broadband provider

Forty-five minutes to find the number in the first place because your internet is down, followed by a chat with a nice young Indian man in Pune who confirms that your internet is down.

•

Mobile phone company

A forty-five minute wait for support. A one-minute wait for sales.

•

Any helpline with a countdown

A robotic voice explains that all the helpline's operators are busy then tells you the number of people ahead of you in the queue. Time it takes to get to the head of the queue: forty-five minutes.

•

Satellite TV

An automated maze of numbers, sub-directories and messaging. You will only speak to a person if the cleaner picks up the call by mistake.

Top 10 most annoying pieces of packaging

Moulded plastic dome around fruit

Is there any greater affront to nature than a single orange encased in a PVC biosphere which requires a buzz saw to open?

•

Individual milk or creamer pot

Even in the hands of seasoned hotel residents these UHT time bombs prove to be dangerously explosive.

•

Parmesan wedge

Peeling from the pointy end as instructed simply lodges the fat end in like an immovable cheese anchor.

•

Giant polystyrene mouldings

They may have kept your new telly safe but now they refuse to fit back into their box of origin.

•

Plastic milk containers

Why does anyone need a screw-top lid and a foil bit?

•

Tortilla chips

The eternal question ... why such a big bag for so few chips?

•

Any Apple product

Looks great but I can't turn it on. Where the hell are the instructions?

•

Children's toys

The devil's own mix of sticky tape, twisty tags, staples and industrially-bonded cardboard means you will need the Jaws of Life to get through Christmas morning.

•

Battery packs

A slit in the back to extract one battery is handy. But there are four in the pack.

•

Sauce sachets

Cut them, bite them, stamp on them. You will still end up with more sauce on your fingers than on the food.

Top 10 ringtones of all-time

Listening to other people wittering away on their mobile phones is annoying, but it's nothing compared to enduring the personalised ringtones that announce the arrival of such moronic exchanges. For people who wanted to express their personality through a tiny, polyphonic announcement, personalised ringtones were a dream come true and a huge industry grew up around selling them. Here are the top 10 mobile ringtones of all-time.

•

'In Da Club' by 50 Cent

A head-bobbing tune for teens too young to set foot in an actual nightclub. It was Billboard's inaugural 'Ringtone of the Year' in 2004.

•

'Stairway to Heaven' by Led Zeppelin

A forty-year-old song that remains enduringly popular with 40-year-old IT consultants.

•

'Sweet Home Alabama' by Lynyrd Skynyrd

Wannabe redneck classic for people who own a pick-up truck but have never left Norfolk.

•

'Super Mario Bros. Theme'

A retro classic for serious gamers. The ringtone is often an exact replication of the track used by the 16-bit Nintendo system, if you're interested.

•

'Halloween Theme'

Used by creepy horror fans and travelling photocopier salesmen who dream of kidnapping a hotel maid and driving her around in the boot of their Ford Mondeo.

•

'The Next Episode' by Dr Dre

Dre's thumping bass and G-funk guitar redesigned to bring the original gangsta feel to the pay-as-you-go phones of a generation of white middle class students.

•

'Theme from Mission: Impossible'

Usually the Lalo Schifrin version from the original series. Adds false urgency to a visit to the dentist.

•

'Sandstorm' by Darude

Popular with people damaged by the drugs they took while listening to techno in their twenties.

•

'The Theme to 'The Fresh Prince of Bel-Air'

A hopelessly upbeat ringtone favoured by people close to a nervous breakdown.

•

'The Jetsons Theme'

Left field choice for people who believe their mobile represents their passport to a technologically advanced future in which they also own a jet pack.

> **'If Bill Gates had a penny for every time I reboot my computer –**
> **Oh wait! he does.'**
> Author unkown

Top 6 ridiculous Wi-Fi hotspots

For chronic internet addicts, the search for Wi-Fi hotspots is an obsession. This lust for connectivity came to a head at the SXSW music and technology conference in Austin, Texas when homeless people were turned into walking Wi-Fi hotspots by the branding agency BBH. The unfortunate vagrants were employed to stand next to conference-goers to provide internet access via Mi-Fi devices connected to 4G phones. The service cost two dollars for fifteen minutes online. Here are five more preposterous Wi-Fi hotspots.

•

Phone booths in Moscow

Moscow has two hundred coin-operated telephones with wireless connectivity to its Comstar network. This means the city's residents face a long wait behind someone playing fantasy football if they need to make a call.

•

Florida State University

Golf course has free Wi-Fi access for people who should probably be concentrating on their game.

•

The subway in Buenos Aires

Argentina became the first city to offer free internet access underground. Now passengers miss their stop while checking their email.

•

Whole Foods Market stores in the US

The chain allows its customers to stay connected while shopping for local food, enjoying an organic soy latte and wearing sandals in public.

•

Marinas in Auckland

New Zealand's Westhaven Marina is now the largest marina Wi-Fi zone in the southern hemisphere. Good news for those who are keen to update their Facebook status to 'On a boat'.

•

Village of Sarohan

The isolated village of Sarohan, with a population of just two thousand people and no mains electricity, has outstanding Wi-Fi connectivity, and is something of a miracle.

Top 5 failed technological predictions

You can probably remember a time when you confidently asserted to your friends that you would never need a mobile phone. Now you're more likely to leave one of your kids at the supermarket than forget to leave the house without your Nokia. But don't beat yourself up about this. The application and power of technology has grown exponentially over the last 100 years and the only thing that has been able to keep pace with it is the ability of those who should know better to write it off. Here are five such examples:

•

'This "telephone" has too many shortcomings to be seriously considered as a means of communication. The device is inherently of no value to us.'

Western Union internal memo from 1876

•

'The wireless music box has no imaginable commercial value. Who would pay for a message sent to nobody in particular?'

Advice given to David Sarnoff, pioneer of television and radio in the 1920s.

•

'I think there is a world market for maybe five computers.'

Thomas Watson, chairman of IBM, in 1943

•

'Computers in the future may weigh no more than 1.5 tons.'

Reported by Popular Mechanics *magazine in 1949*

•

'There is no reason anyone would want a computer in their home.'

Ken Olson, president, chairman and founder of Digital Equipment Corp, in 1977

•

And finally some words of wisdom from a man who REALLY should know better:

'Nobody will ever need more than 640k of RAM!'

Bill Gates, founder of Microsoft and philanthropist in 1981

•

Closely followed by ...

'Windows 95 needs at least 8 MB of RAM.'

Bill Gates, founder of Microsoft and philanthropist in 1996

Premium food brands

Thanks to some inspired yet essentially evil work by branding gurus, you now believe that you have to buy premium products to treat yourself and – more importantly – to show your nearest and dearest that you're not a cheapskate, who would rather have them eat their own shoe leather than fork any extra couple of quid on orange juice with a nice illustration of Seville on the carton.

Indeed, nowadays you wouldn't dare to serve anything but a premium brand of ice cream when people come to dinner, despite the fact that no one can tell the difference between a sumptuous carton of Madagascan Vanilla and a block of melted hooves dyed white and wrapped in cardboard in a grimy factory.

The real kicker is that as soon as you buy one of these premium brands you can never again save money on the value brands. Despite being fully aware that even the highest quality sparkling water is probably just drawn from the tap and fired through a soda stream, you can't pop a bottle of value sparkling water in your cart without feeling like everyone in the queue imagines that you've recently been made redundant.

'A bargain is something you can't use at a price you can't resist.'
Franklin P. Jones

Top 10 bestselling 'Finest' brand products

A retail analyst said of Finest and Value ranges: 'The supermarkets have pulled off a trick that I'm not aware of any other retail sector achieving. That is to appeal to all segments of the market.' What he meant by that was that this perceived variance in quality and actual variance in price created by the retailer has made every trip you take to the supermarket a guilt-fuelled journey to centre of your own soul.

Here is a more realistic view of their bestsellers

•

Piccolo Cherry Tomatoes

Smaller than normal tomatoes but they've left the stalks on for that back to nature feel.

•

Wafer Thin Cured Ham

Sliced pig's arse with a little rim of breadcrumbs on it. Cut so finely you have to put two slices in to make a sandwich. You do the maths.

•

Steak Beef Burgers

Allegedly made from the "finest beef" for those who were unaware that both steak and beef come from cows.

•

Free Range Corn Fed Chicken

A yellow chicken that was raised in a loving and nurturing environment before being decapitated and disemboweled by an automated plucking machine.

•

Pinot Grigio

White wine that French people probably wouldn't brush their teeth in. The label looks posh simply by not being the name of a minor geographical feature in the antipodes (eg Warren's Cliff).

•

Sirloin Steak

Sold on the fact that it supports local farmers – especially if they want to convert their farm to producing meat exclusively for a single supermarket chain.

•

14 Month Genuine Parma Ham

The antidote to black market Chinese Parma Ham being sold by illegal hawkers nearby.

•

Multi-grain Batch bread

By batch they mean baked in batches of 25,000 in windowless warehouses staffed by migrant workers.

•

Butter Roast Turkey

It's a turkey with a nob of butter on top. Why didn't you think of that?

•

Frozen Jumbo King Prawns

Big prawns.

Top 5 crazy shoppers

We all like a bargain but how far would you go to get one? In the US the first shopping day after Thanksgiving is traditionally known as Black Friday. It is called this because it is the first day of the autumn sales season and helps retailers into the black for the first time in the financial year – unofficially it has become the symbol of consumerism at its most crazed, as shoppers riot to secure bargains. Here are just a few incidents from 2011:

•

In New Jersey, a woman repeatedly stabbed a man with a kitchen knife taken from the homeware section of Wal-Mart to prevent him from buying the store's last Xbox 360.

In Columbus, Ohio, a shortage of *Star Wars* toys created a stampede in which four people were trampled after someone yelled, 'they've got some over here!'

●

In Tucker, Georgia, a woman bludgeoned a fellow shopper to death with a baseball bat over an American Girl doll in Toys R Us while yelling 'hasta la vista baby'.

●

In Los Angeles, California, a Hispanic woman pepper-sprayed a crowd as they fought over a fresh consignment of Xboxes, before making good her escape.

●

In Myrtle Beach, South Carolina, a 55 year old woman was shot in the foot by a robber as she loaded hard earned bargains into the boot of her car. Luckily her companions were also armed and the robber ran off when they returned fire.

Top 5 Facebook updates that went badly wrong

For most of us the most creative we'll ever get with our Facebook status is changing it from 'In a relationship' to 'It's complicated' following an argument with our partner over who's turn it is to put out the recycling. However others have taken this public display of their innermost feelings to a new level and ultimately ended up coming a cropper. Here are five examples of people who really should have worn their heart on their sleeve rather than on their PC.

•

Keeley Houghton had terrorized her neighbour Emily Moore for four years but it was only when she posted a death-threat against Moore on Facebook that she became the first person in the UK to be jailed for internet bullying and served 3 months in jail.

•

When Craig 'Lazie' Lynch escaped from the prison where he was serving a seven-year stretch for armed robbery, he posted a series of images of him giving the finger to the authorities on Facebook (in some cases literally) instead of going to ground. His enthusiastic disdain for authority meant he collected over 40,000 fans and inspired a tribute song before he was recaptured and returned to jail.

•

'Unemployed single mum' Hazel Cunningham was happily claiming income support, housing benefit and council tax benefit, when government investigators noticed pictures of her and her children at her wedding (in Barbados) and various family holidays in Turkey on Facebook. She received a 120-day prison sentence and had to pay back £15,000.

•

Inspired by the London riots friends Perry Sutcliffe-Keenan and Jordan Blackshaw set up a Facebook page which invited everyone to help them 'Smash down in Northwich Town.' Unfortunately for the hapless teenagers only the police took them up on the offer and they were sentenced to four years in jail.

•

Proud hunter, Brandon Lowry of Norco, Louisiana, posted pictures of himself with a grisly haul of 64 ducks he had despatched during a single hunting trip. Unfortunately the maximum amount hunters can kill in the state during teal season is eight, and The Louisiana Department of Wildlife and Fisheries felt a fitting tribute to Lowry's skills with a rifle was a $950 fine and 120 days in jail.

Top 5 movie trailers

There is nothing quite as exciting, nor as misleading as a movie trailer. Whether you are slumped in the cinema waiting for the main feature or flicking between TV channels, you consume this fast-cut beauty with a rising pulse before making a mental note to book a babysitter for the day of the film's release.

Funny business

In the cold light of day, these single-minded plugathons are unhelpful as they make every film look fantastic regardless of its merit. Whether it is a harrowing true story set in a concentration camp with Ian McKellen, a knockabout comedy with Eddie Murphy, a romance with Brad Pitt or a summer blockbuster involving the computer-generated destruction of a major urban landmark, the trailer promises so much.

The disappointment comes when you get to the movie theatre. After driving five miles out of town to a cinema cum bowling alley cum fast food outlet the size of a small airport, you dutifully queue up behind a hoard of spotty teenagers to hand over the price of a dinner for two at a modest restaurant in return for a pair of tickets, a bucket of cola and a box of chemically-enhanced popcorn. Much of this will be spilt as you walk the five miles to Screen 78 and fumble through the darkness and deafening Dolby surround sound to find your heavily-stained pre-booked gondola. There you will sit through another five trailers (which also look brilliant) before discovering that the film you came to see is a load of old rubbish. To add insult to injury, you've all ready seen all the good bits, heard all the best jokes and know the plot from the trailer.

The bottom line is that the trailer is so much better than the movie itself. There was no need to come to the cinema at all. In future, it might make sense to cut out the middle man and simply give the kick-ass trailer a cinematic release and leave the disappointing full-length movie to be shown on YouTube as a teaser.

The Blair Witch Project

Only 40 seconds but absolutely terrifying 'I am so scared', says the girl. So are we dear, so are we.

•

Terminator 2: Judgement Day

Molten people, giant explosions and Arnie appearing through the smoke to drawl, 'Stay here, I'll be back.'

•

Flash Gordon

'Flash aaaaaargh saviour of the universe' – more like a music video than a film trailer. Words by Ming the Merciless, musical pomp by Queen.

•

Independence Day

All the extra special effects from the movie in an easy-to-digest two minute portion.

•

Psycho

Alfred Hitchock provides a tour guide round the Bates Motel as it was real.

Also from Michael O'Mara Books

Universally Challenged
Wendy Roby
£7.99
ISBN-13: 978-1-84317-466-0

Also from Michael O'Mara Books

Are We Live?
Marion Appleby
£7.99
ISBN-13: 978-1-84317-866-8

Also from Michael O'Mara Books

S*** My Kids Say
Nick Harris
£7.99
ISBN-13: 978-1-84317-867-5

Also from Michael O'Mara Books

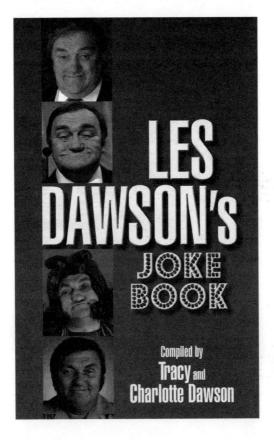

Les Dawson's Joke Book
Compiled by Tracy and Charlotte Dawson
£9.99
ISBN-13: 978-1-84317-870-5

Also from Michael O'Mara Books

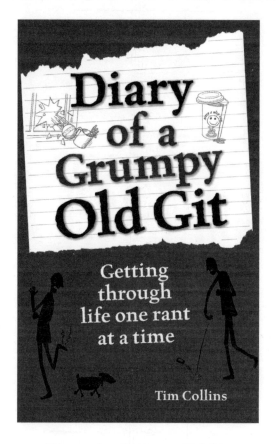

Diary of a Grumpy Old Git
Tim Collins
£9.99
ISBN-13: 978-1-84317-949-8